Numbers for Smart Kids

Betsy Wolff Frey

Contents

Introduction

This book is for students who want to learn more about numbers and how they are used. It introduces the reader to a world beyond simple arithmetic, providing an overview of the broad world of mathematics.

If a chapter is too simple, feel free to skip it. If a chapter is too complicated, don't worry about it. Some of the later chapters deal with complex issues that are often covered at the college level.

The book can be used by students who wish to review the math they have already learned, and to peek into areas of mathematics they have not yet studied. The book can be used by parents who are home-schooling their children. The book can also be read by old folks who once knew what a logarithm was, but would appreciate a refresher course.

Welcome to the world of numbers.

Math Problems? Call $1 - 800 - 3x^3 - 12\pi e^x$.

Notation

Note that letters such as *c* and *i* do not always have the special meaning listed here. Like other letters, they sometimes represent a constant or variable in an equation.

Symbol	Name	Page	Meaning
a, b, ...	a, b, ...		constant or variable in an equation
$\aleph$	aleph	82	an infinite cardinal number
$\approx$			symbol for *approximately equal to*
c	c	41	the speed of light
e	e	42	Euler's number, about 2.71
i	i	46	the imaginary unit, $\sqrt{-1}$
∞	infinity	6	infinity
$\int$	integral	74	the integration symbol
ϕ	phi	44	the golden ratio, about 1.6
π	pi	35	$\frac{\text{circumference}}{\text{diameter}}$ of a circle, about 3.14
$\sqrt{}$	radical	30	a root of a number
Σ	sigma	58	the summation symbol.
θ	theta	63	a variable angle measurement

Chapter 1

Kinds of Numbers

1.1 Counting Numbers

The counting numbers are 1, 2, 3, ... Counting begins with one.
The counting numbers continue on forever. This is known as con-
tinuing into *infinity*.

Question: Why was 6 afraid of 7?
Answer: Because 7 8 9.

Every counting number is interesting.

Proof: Assume there was an uninteresting counting number.
Then there must be a *smallest* uninteresting counting number.
Being the smallest uninteresting counting number is interesting.
This is a contradiction.
Therefore, every counting number is interesting.

1.2   Negative Numbers

Negative numbers are less than zero. Each positive number has an equivalent negative number on the number line.

Table 1.1: The Number Line

...	-4	-3	-2	-1	0	1	2	3	4	...

Question: If you are 8 years old, how old were you 10 years ago? *Answer:* Huh? Well, mathematically speaking, I suppose I was negative 2 years old. Also known as -2.

In business accounting, positive numbers represent money the business has earned. Negative numbers are used for money the business owes. Add the numbers to get the total. That is called *the bottom line*. Be happy if the bottom line is positive.

Amount	Reason
+100	in bank account
+1,400	owed to us for crops shipped out
-300	payment due for tractor
-200	payment due for seed
-950	owed to farm workers
+50	bottom line

1.3 Integers

Integers are the counting numbers and their negatives and zero. Integers are the whole numbers on the number line.

1.4 Fractions

A *fraction* is a ratio of two numbers. It is equivalent to the first (top) number, called the *numerator*, divided by the second (bottom) number, called the *denominator*. The bottom number cannot be zero, because division by zero is not allowed.

Think of a pizza that has been cut into slices of equal size. The numerator defines the number of slices, while the denominator defines the number of slices in one whole pizza. The pizza might have been divided into 2, 4, or 20 slices. The fractions $\frac{1}{2}$, $\frac{2}{4}$, $\frac{10}{20}$ are equivalent: each represents half the pizza.

A fraction has been *normalized* if there is no equivalent fraction with a smaller denominator. The fraction $\frac{1}{2}$ is the normalized form of $\frac{10}{20}$.

1.5   Rational Numbers

Rational numbers can be expressed as a fraction $\frac{x}{y}$ of two integers x and y, where y is not zero.

Rational Number	Fraction	Decimal Notation
seven hundred	$\frac{700}{1}$	700
zero	$\frac{0}{1}$	0
two and a half	$\frac{5}{2}$	2.5
one quarter	$\frac{1}{4}$	.25
three tenths	$\frac{3}{10}$	.3

Some rational numbers go on forever when written in decimal notation. For example, $\frac{1}{3}$ of a dollar is a little more than 33 cents. $\frac{1}{3}$ is the same as:

$$0.33333333333333333333$$

where the 3's keep going on forever.

1.6 Irrational Numbers

Irrational people are people who are not logical.

Irrational numbers behave logically. They are called *irrational* because they are not rational numbers; they cannot be expressed as a fraction of two integers.

When an irrational number is written in decimal notation, it goes on forever, like this: 5.1423957123250573845... Some rational numbers, like $\frac{1}{3}$ (0.333333...), go on forever; but *all* irrational numbers go on forever.

An irrational number cannot be written in decimal notation, because it goes on forever. It cannot be written as a fraction of two integers, because then it would be a rational number. Names have been assigned to several interesting irrational numbers. This book discusses some special irrational numbers.

Irrational Number	Name	Approximate Value
$\sqrt{2}$	the square root of 2	1.41421
π	pi	3.14159
e	Euler's number	2.71828
ϕ	phi	1.61803

1.7   Real Numbers

Real numbers are all the rational and irrational numbers. They
include all the integers and all the points between any two adjacent
integers X and X+1. The set of real numbers is the set of all the
points on the number line.

Table 1.2: Points on the Number Line

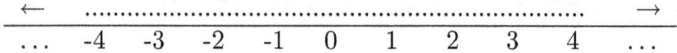

		-4	-3	-2	-1	0	1	2	3	4	...

The term *real number* might sound strange — how can a number
not be real?

Keep reading.

1.8   Infinite Numbers

Infinity is not a real number. Positive infinity $+\infty$ is the end of the
positive integers. Negative infinity $-\infty$ is the end of the negative
integers. There are other infinite numbers. These are discussed in
section 7.8 *Transfinite Arithmetic* on page 81.

1.9 Complex, Imaginary Numbers

Complex numbers have two parts: a real part and an *imaginary* part. They have the form:

$$a + bi$$

where a and b are real numbers, and i is the *imaginary unit*. The imaginary unit has an odd name, but is a very useful mathematical concept. It is defined as *the square root of -1*, which means:

$$i \times i = -1$$

It is discussed in section 5.5 *i, the Imaginary Unit* on page 46.

Fact: Life is complex. It has real and imaginary components.

Chapter 2

Notation

Suppose you have some arrows.

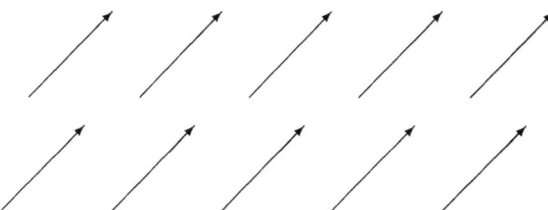

How many arrows are there? Well, that depends...

2.1 Decimal Notation, base 10

You have *10* arrows, in *decimal positional notation*. This notation is also called *base ten*. Decimal positional notation is the way we usually write numbers, but one can write numbers in other ways, such as base two or base sixteen.

> *Question:* Why is it that our number system is base 10?
> *Answer:* Count your fingers.

In positional notation, the columns to the left of the decimal point are used for bigger and bigger values—in decimal notation, the columns are for units, tens, hundreds, thousands, etc. The columns to the right of the decimal point are used for smaller and smaller values—in decimal notation, the columns are for tenths, hundredths, thousandths, and so on.

Our money system is based on our use of decimal notation. Except for nickels and quarters and fifty cent pieces. And five dollar bills and twenty dollar bills.

Correction: our money system is *partly* based on our use of decimal notation.

Table 2.1: Columns for $123.45 in Decimal Notation

$1000's	$100's	$10's	$1's		dimes	pennies
	1	2	3	.	4	5

Other Bases

Positional notation always has a base. Usually the base is ten, which is called decimal. For other bases, a small subscript number gives the value of the base. 10_2 is pronounced *ten, base two.*

The smallest allowable base is two. Just as base ten uses the digits 0–9, any base b must have a unique digit for every value between 0 and $b - 1$. Since the smallest allowable base is 2, every base has the digits 0 and 1. For every base b:

the column immediately left of the decimal point represents units

the column left of that represents b's.

the column left of that represents $b \times b$'s.

the column left of that represents $b \times b \times b$'s, etc.

$1_b = 1$

$10_b = b$

$100_b = b \times b$

$1000_b = b \times b \times b$

2.2   Binary Notation, base 2

Remember the arrows? You did not get any more arrows, even though you have $1,010_2$ arrows. The small number 2 tells us the number is written in base 2. Base 2 is also known as *binary*. Binary notation uses only two digits: 0 and 1.

Decimal Dan and Binary Betty count the arrows differently.

Decimal Dan	Binary Betty
1	1
2	10
3	11
4	100
5	101
6	110
7	111
8	1000
9	1001
10	1010

There are 10 kinds of people: those who can think in binary and those who can't.

$$10 = 10_{10} = 1010_2$$

The number 10 in decimal is the same as the number 1010 in binary. That is because $10 = 8 + 2$. Binary notation has columns for units, twos, fours (2 x 2), eights (2 x 2 x 2), and so on.

Table 2.2: 10_{10} in Binary Notation

128's	64's	32's	16's	8's	4's	2's	units
				1	0	1	0

Question: How old are you in binary?

2.3   Hexadecimal Notation, base 16

Remember the arrows? Now you have A_{16} arrows. The small number 16 tells us the number is written in base 16.

Base 16 is also known as *hexadecimal*, sometimes shortened to *hex*. Hexadecimal uses the digits 0–9, but it needs 6 more digits, for the values 10–15. It uses the letters A-F to represent the values 10-15.

$A = 10$	$D = 13$
$B = 11$	$E = 14$
$C = 12$	$F = 15$

Tex Hex has his own way of counting the arrows.

Decimal Dan	Binary Betty	Tex Hex
1	1	1
2	10	2
3	11	3
4	100	4
5	101	5
6	110	6
7	111	7
8	1000	8
9	1001	9
10	1010	A

Hexadecimal notation has columns for units, 16's, 256's (16 x 16), 4096's (16 x 16 x 16), and so on.

$$10 = 10_{10} = 1010_2 = A_{16}$$

The number 10 in decimal is the same as the number A in hex. The number 1,000 in decimal is the same as the number 3E8 in hex. That is because:

$$1,000 = 1000_{10} = (3 \times 256) + (14 \times 16) + 8$$

Table 2.3: 1000_{10} in Hexadecimal Notation

65536's $16 \cdot 16 \cdot 16 \cdot 16$	4096's $16 \cdot 16 \cdot 16$	256's $16 \cdot 16$	16's	units
		3	E	8

Question: What's BAD_{16}?
Answer: 2,989, which is $(11 \times 256) + (10 \times 16) + 13$

Computer Storage

People who work with computers sometimes use hex notation. Computers use binary values. The basic unit of computer storage is a *bit*, which has a value that is either zero or one. Since 4 binary digits represent a value between 0 and 15, every 4 binary digits can be converted to one hexadecimal digit.

Working with binary numbers can make a person cross-eyed. It is enormously easier for a person to work with a 4 digit hex value like $A299$ than a 16 digit binary value like 1010001010011001.

Eight bits is called one *byte*. A byte can hold 256 different values, from 00_{16} to FF_{16}. Sometimes a byte is called a *character*, because it is enough to represent any one of the characters on a keyboard: a capital letter A–Z, or a lower-case letter a–z, the digits 0–9, and many different punctuation marks.

Table 2.4: Selected ASCII Character Codes

Hexadecimal Byte Value	Character
20	[space]
2E	.
31	1
39	9
3A	:
41	A
5A	Z
61	a
7A	z

Many new words have appeared because of computers. A *gigabyte* is $1,000,000,000$ (one billion) bytes. A byte can hold one character. A gigabyte can hold all the words in an encyclopedia.[1]

value	abbreviation	# bytes	# bytes
bit		1/8	
byte		1	one
kilobyte	KB	1,000	thousand
megabyte	MB	1,000,000	million
gigabyte	GB	1,000,000,000	billion
terabyte	TB	1,000,000,000,000	trillion
petabyte	PB	10^{15}	quadrillion
exabyte	EB	10^{18}	quintillion
zettabyte	ZB	10^{21}	sextillion
yottabyte	YB	10^{24}	septillion

---

[1] The Enclyclopedia Britannica has about 40,000,000 words. If on average there are 5 letters per word, plus a blank to separate words, then that would be 240,000,000 characters. So a gigabyte could actually hold more than 4 copies of the words in the Enclyopedia Britannica.

2.4   Sexagesimal Notation, base 60

Remember the arrows? If you were a Babylonian living 5,000 years
ago, you would have ◁ arrows. The Babylonians used a base 60
number system. They had different symbols for the numbers 1
through 59. Later, they added a symbol for zero.

Base 60 is also known as *sexagesimal*. The Babylonians cut their
symbols into stone tablets. Three of their symbols look something
like this:

$$\bowtie = 0$$
$$\bar{Y} = 1$$
$$\triangleleft = 10$$

Sexagesimal notation has columns for units, 60's, 3600's (60 × 60),
and so on.

$$10 = 10_{10} = \triangleleft$$

The Babylonians did not have a decimal point. When they wrote a
number between zero and one, they began the number with a zero.
They would write one second as ⋈ ⋈ Ȳ.

Table 2.5: One Second, in Babylonian

hours	.	minutes ($\frac{1}{60}$ hour)	seconds ($\frac{1}{3600}$ hour)
	⋈	⋈	Ȳ

We measure time like the Babylonians did. This is why there are
60 seconds in a minute and 60 minutes in an hour.

2.5 Roman Numerals

Remember the arrows? If you were a Roman living 2,000 years ago, you would have X arrows. The Romans did not use a positional number system, instead they had these symbols.

$$I = 1 \qquad L = 50 \qquad M = 1000$$
$$V = 5 \qquad C = 100$$
$$X = 10 \qquad D = 500$$

There is no zero in the system of Roman numerals. Because it is not positional, there is no need to specify a column value of zero.

Regina Romana has her own way of counting the arrows.

Decimal Dan	Binary Betty	Tex Hex	Regina Romana
1	1	1	I
2	10	2	II
3	11	3	III
4	100	4	IV
5	101	5	V
6	110	6	VI
7	111	7	VII
8	1000	8	VIII
9	1001	9	IX
10	1010	A	X

These are some of the rules for Roman numerals:

1. Write values from left to right, greatest values first; e.g., CI means 101.

2. If a smaller value symbol is to the left of a larger value symbol, subtract it from the larger value symbol. For example, IV means 4.

3. It is not correct to use more than one smaller value symbol to the left of the larger one. For example, IIX is not valid.

4. It is not correct if the smaller value symbol to the left of a larger value symbol is more than 1 or 2 levels smaller than the larger symbol. For example, IL is incorrect and does not mean 49. This is 49 (40 + 9): XLIX.

Table 2.6: Some Roman Numeral Values

$11 = XI$	$16 = XVI$	$77 = LXXVII$
$12 = XII$	$17 = XVII$	$88 = LXXXVIII$
$13 = XIII$	$18 = XVIII$	$99 = XCIX$
$14 = XIV$	$19 = XIX$	$601 = DCI$
$15 = XV$	$20 = XX$	$2009 = MMIX$

2.6   Spoken Numbers

Although we use positional notation when writing the number 2009, we do not pronounce it *two zero zero nine*; instead, we say *two thousand nine*.

Languages sometimes have spoken forms that differ from their written forms; this is an example of the way they can differ. The way we pronounce 1000 (one thousand) is more like the way the Romans wrote it (one M) than the way we write it.

Are there *one zero* arrows? Don't be silly. There are *ten* arrows.

2.7 Notation Summary

How many arrows are there? What is the right answer?

The answer depends on the context.

CONTEXT	ANSWER
decimal	10
binary	1010
hexadecimal	A
Babylonian	◁
Roman	X
spoken	ten

Chapter 3

Bigness and Smallness

3.1 The Biggest Number

Question: What is the biggest number?
Answer: When asked this question, children have responded:

- two thousand and four aces
- 10 tillion
- 10 billion and 10 thousand and 10 hundred
- oh... what's that word... the word that means numbers go on and on

The word is *infinity.* Infinity is a pretty good answer to the question.

There is a problem, though: there are several different infinities, and some are bigger than others. See section 7.8 *Transfinite Arithmetic* on page 81.

3.2 The Smallest Number

Question: What is the smallest number?
Answer: That is an interesting question.

The smallest counting number is 1.

A philosopher once argued that the smallest number was 2. His thinking was that One is not really a number; it is the idea of Unity, Being Whole. He said counting does not start until you get to 2.

Other people might say the smallest number is 0. Anything is more than nothing. Is zero a number, or is it just the idea of nothing? The Babylonians used numbers for more than a thousand years before they invented a symbol for zero. The Romans did not have a symbol for zero.

It is currently thought that there is a smallest length about which anything can be known. It is impossible to measure a size smaller than one *Planck length*, which is about 6.3×10^{-34} inches.:

$$0.00000000000000000000000000000000063 \text{ inches}$$

Similarly, it is impossible to measure a time interval smaller than the time it takes light to travel one Planck length. A *Planck time* is about 5×10^{-44} seconds:

$$0.00000000000000000000000000000000000000000005 \text{ seconds}$$

Even if one cannot measure smaller than one Planck length or one Planck time, one can still think about numbers smaller than those values.

Consider negative numbers: -1 is less than zero, and -2 is less than -1. Does that mean the smallest number is -9999999999...? That seems like a very big number, even though it is negative.

What *is* the smallest number? Good question.

Chapter 4

Numeric Operations

4.1 Arithmetic

Arithmetic consists of addition, subtraction, multiplication, and division. Arithmetic is an important part of mathematics, but mathematics is much more than arithmetic. Some mathematicians are not very good at arithmetic, because it is easy to make mistakes. Computers are excellent at arithmetic.

People often question the value of learning about higher mathematics, like trigonometry or calculus. Very few people question the value of learning arithmetic.

Arithmetic is definitely useful in life; however, arithmetic is boring.

Addition

Addition is adding numbers. If you do not understand addition, you should probably stop reading this book.

Subtraction

Subtraction is a form of addition which uses negative numbers. $8 - 6$ is the same as $8 + (-6)$.

Multiplication

Multiplication is super addition. 8×6, which is *8 times 6*, means *adding together six 8's*: $8 + 8 + 8 + 8 + 8 + 8 = 48$.

When a symbol represents a number, one can skip the $\times$ symbol: $4A$ means $4 \times A$, and AB means $A \times B$. Sometimes the $\cdot$ symbol is used instead of the $\times$ symbol; each means multiplication.

When multiplying a positive and a negative number, the result is always negative; e.g., $-3 \cdot 4 = -12$ and $3 \cdot -4 = -12$.

When multiplying two negative numbers, the result is always positive; e.g., $-3 \times -4 = 12$. Please don't ask why.

Division

Division is like reverse multiplication, because if A times B is C, then C divided by B is A.

For example, $12/4 = 3$. This can also be written $12 \div 4 = 3$ or $\frac{12}{4} = 3$. It means *12 divided by 4 is 3*. It also means that if one is going to divide 12 cupcakes among 4 hungry people, then each person gets 3 cupcakes. I like that.

THIS IS THE END OF THE BORING SECTION
ABOUT ARITHMETIC.

4.2   Exponents

Positive Integer Exponents

Exponents are a way of expressing multiplication of a number times itself. For example,

$5^2 = 5 \times 5$. This is also called 5 squared or 5 to the power of 2.

$2^3 = 2 \times 2 \times 2$. This is also called 2 cubed or 2 to the power of 3.

$3^7 = 3 \times 3 \times 3 \times 3 \times 3 \times 3 \times 3$. This is called 3 to the power of 7.

Squared means raised to the exponent (or power) of 2. *Cubed* means raised to the exponent (or power) of 3.

A *Handy Equation:*

$$b^i \times b^j = b^{i+j}$$

Think of the value of $(b^3) \times (b^4)$.

$$(b^3)(b^4) = (bbb)(bbbb) = b^7$$

Other Exponents

There are mathematical definitions for other exponents.

zero exponent: any number to the power zero is always 1. This keeps the Handy Equation true. Think of the rule that

$$x^n = (x^{n+1} \div x)$$

That means that $x^0 = x \div x = 1$.

negative exponent: x to the power -y, where y is a positive integer, is $\frac{1}{x^y}$. For example,

$$10^{-3} = \frac{1}{10^3} = \frac{1}{1000} = 0.001$$

fractional exponent: x to the power $\frac{m}{y}$, where x is a non-negative number and m and y are counting numbers, is the y'th root of x^m. Roots are discussed in the next section.

imaginary exponent: Raising a number to a complex power gets complicated. The next section discusses the fact that there are two values for the square root of a positive real number. Similarly, there can be many different values for z^c when both z and c are complex.

4.3   Roots

The *square root* of a number X is the number Y, where $Y \times Y = X$.
Every positive number has two square roots: one is positive and
the other is negative. For example, the square roots of 4 are 2 and
-2, because $2 \times 2 = 4$ and $-2 \times -2 = 4$. The square roots of 49 are
7 and -7.

A square root is shown using a radical symbol, like this:

$$\sqrt{49} = 7$$

A square root can also be shown using exponent notation, with an
exponent of $\frac{1}{2}$:

$$49^{\frac{1}{2}} = 7$$

There is no real number that is the square root of a negative number, because any real number times itself is either zero or positive.

There are other kinds of roots. The *cube root* of a number X is the
number Y, where $Y \times Y \times Y = X$. For example, $\sqrt[3]{8} = 2$, because
$2 \times 2 \times 2 = 8$.

The following are equivalent ways of stating that Y is the *nth root*
of a number X:

$$\sqrt[n]{X} = Y \qquad\qquad X^{\frac{1}{n}} = Y \qquad\qquad Y^n = X$$

Square Root of Two

Like the roots of most numbers, $\sqrt{2}$ is an irrational number. These
are the first 199 digits in the positive square root of 2:

```
1.4142135623730950488016887242096980785696716718675376
 9480731766797379907324784621070388503875343276415
 273501384623091229702492483605585073721264412149970
 9993583141322266592750559275579995050115278206057
```

The *area* of any square whose side is length S is $S \times S$, so a square whose area is one has sides of length one. What is the length of the diagonal in a square whose area is one?

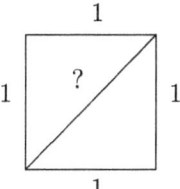

Take two of the above squares, cut them on the diagonal, and form them into a new square. Because the new square is made from two squares having an area of one, the area of the new square is two. That means the new square has sides whose length is $\sqrt{2}$, the square root of two.

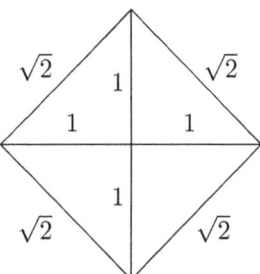

Hence, a square whose area is 1 has a diagonal whose length is $\sqrt{2}$.

4.4   Logarithms

Logarithms relate to a base and an exponent. If $x = b^y$, then $log_b(x) = y$. This says that *the logarithm of x to the base b is y*.

Common Logarithms

Common logarithms are logarithms to the base 10. The common logarithm of 1000 is 3: $log_{10}(1000) = 3$, because $10^3 = 1000$.

Table 4.1: Some Common Logarithms

value	0.01	0.1	1	10	100	1,000	10,000	100,000
log(value)	-2	-1	0	1	2	3	4	5

When a value changes a lot, its logarithm changes just a little. Common logarithms are often used for measurements that have a wide range of values:

- the **Richter scale** measures the strength of earthquakes. An earthquake that measures 6.0 on the Richter scale is ten times stronger than one that measures 5.0.

- **decibels** measure loudness. A decibel is one tenth of a bel[1]. 150 decibels is the same as 15 bels, which is about the noise of a jet engine. 150 decibels is 100,000,000,000,000 times louder than a quiet whisper of 10 decibels, which is 1 bel.

- **pH** measures the acidity of liquids. Water, which is neutral, has a pH of 7.0. Liquids with a pH less than 7 are acidic. Liquids whose pH is greater than 7 are *alkaline*, which is also called *basic*. Lemon juice with a pH of 2 is ten times more acidic than vinegar with a pH of 3.

---
[1] The word *bel* honors Alexander Graham Bell, the inventor of the telephone.

Most numbers have logarithm values that are irrational numbers.

$$log_{10}(2) \approx 0.3010299956639$$
$$log_{10}(26) \approx 1.4149733479708$$
$$log_{10}(200) \approx 2.3010299956639$$
$$log_{10}(2009) \approx 3.3029799367482$$

Base 2 Logarithms

Base 2 logarithms are used in computers and in *cryptology*, which is the study of making and solving secret codes. The logarithm of 16 to the base 2 is 4: $log_2(16) = 4$, because $2^4 = 16$.

Addition instead of Multiplication

Logarithms provide a way of using addition instead of multiplication, because the logarithm of two numbers multiplied together is the sum of the logarithms of each number:

$$log(xy) = log(x) + log(y)$$

Before computers, one could use logarithms to make arithmetic calculations much faster, provided one had two sets of books. These books might be the most boring books ever printed. They contain:

- pre-calculated logarithms, where one can look up a number to get the first few digits of its logarithm.

- pre-calculated antilogarithms, where one can look up a logarithm value to get the number

To get the value of x times y:

1. Look up the value that is the first few digits of the logarithm of x.

2. Look up the value that is the first few digits of the logarithm of y.

3. Add those two numbers together.

4. Look up the result in the *antilogarithm* table, to find the approximate result of x times y. The result is approximate because the books only list the first few digits of logarithm values that are irrational.

For very big numbers x and y, three lookups and some addition is much faster than multiplication without using a computer. Now that we have computers and calculators, tables of logarithms are no longer needed. It is much more convenient to have a calculator in one's pocket than to carry books of logarithmic tables.

After the flood, Noah let all the animals out of the ark. As they left, he told them, "Go forth and multiply."

Two snakes, common European adders, looked at him sadly. "Please," they whispered, "could you chop up a tree for us?"

Noah did as they asked. The snakes smiled. "Why," asked Noah, "did you want the tree to be chopped up?"

"We are adders," the snakes replied. "We need logs to multiply."

Chapter 5

Some Favorite Values

5.1 π, pi

π is a letter in the Greek alphabet, named *pi*, pronounced *pie*.

> *Question:* If you have a circle which is 1 mile long straight
> through the middle, then how long is the path around the
> outside of the circle?
> *Answer:* π miles long.

The path through the middle is known as the *diameter* of the circle. The path around the outside is known as the *circumference* of the circle.

A circle whose diameter is 1 mile has a circumference which is a little more than 3 miles.

To be more precise, it is a little more than 3.1 miles.

To be more precise, it is a little more than 3.1415 miles.

No matter the size of the circle, π is the ratio of the circumference of the circle to its diameter.

$$\pi = \frac{\text{circumference}}{\text{diameter}}$$

The value of π is an irrational number. The first part of the value of π is:

```
3.14159265358979323846264338327950288419716939937510
58209749445923078164062862089986280348253421170679
82148086513282306647093844609550582231725359408128
4811174502841027019385211055596446229489549303081
```

Question: Who cares?
Answer: Farmer Jo cares, at least about the first few digits of π.
A sprinkler system is used to water the back field. The sprinkler is anchored in the middle of the field and sweeps around in a circle.
The sprinkler is 500 feet long, so the circle that gets watered has a diameter of 1,000 feet.
Farmer Jo wants to put a fence around this circle.
How many feet of fencing are needed?
Farmer Jo needs 3,142 feet of fencing.

Any excuse for a party: celebrate Pi Day on March 14 (3.14).

Angle Measurement: Degrees and Radians

An *angle* is formed when two straight lines meet at a point, such as these:

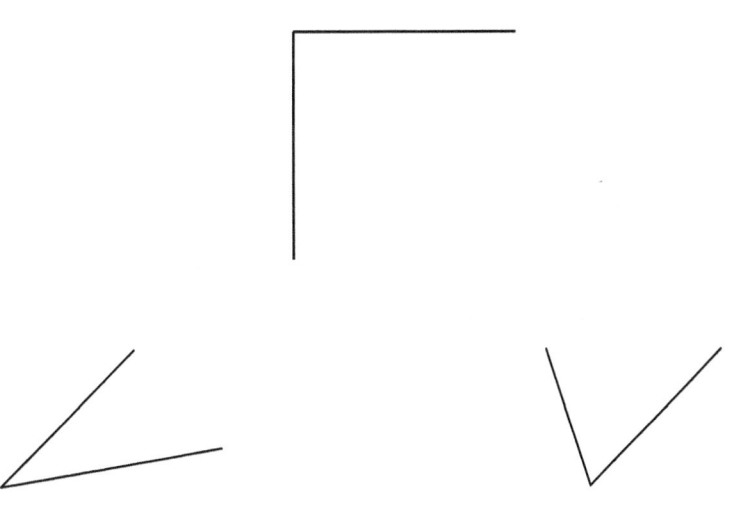

Angles are important in geometry and trigonometry.

When measuring the size of an angle, one is measuring the width of its opening. This is independent of the length and direction of the lines. Angles have two different measurement units:

- *degrees*, thanks to the Babylonians
- *radians*, thanks to π

Angles are often depicted inside a unit circle. Picture a circular clock face with one fixed hand that always points to 3 and another hand that can be moved. The circle is called a *unit* circle because its radius equals one. The *radius* is the length of the hands, half the diameter of the circle.

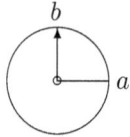 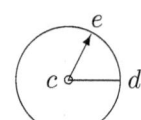

90 *degree angle*   1 *radian angle*
$\pi/2$ *radians* ≈ 57.3 *degrees*
≈ 1.57 *radians*

A circle has 360 equal degrees. Each degree has 60 minutes, and each minute has 60 seconds, compliments of the Babylonians. An angle of 90 degrees, 1 minute, 2 seconds can be written 90° 1' 2". Each quadrant (quarter circle) consists of 90 degrees.

A circle has 2π radians. Each quadrant consists of $\pi/2$ radians. Because the unit circle has a radius of 1 unit, it has a diameter of 2 units; therefore, its circumference is 2π units. Measure the length along the outside edge of the circle between the two lines that form the angle. For a 90° angle, this distance would be $\pi/2$ radians, the distance around the outside from a to b above. For the 1 radian angle above, the curved line from d to e is the same length as the straight lines from c to d and c to e.

$$360 \text{ degrees} = 2\pi \text{ radians}$$
$$1 \text{ degree} = \frac{\pi}{180} \text{ radians} \approx .017 \text{ radians}$$
$$1 \text{ radian} \approx 57.3 \text{ degrees}$$

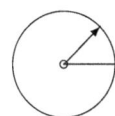

45 degree angle
.25 π radians

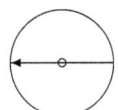

180 degree angle
π radians

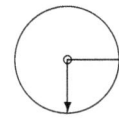

270 degree angle
1.5 π radians

-90 degree angle
-0.5 π radians

Spin one hand of a zero degree angle around in a circle, back to where it began. Spin as many times as you wish. Each spin is 360°. The resulting angle of $N \times 360°$ or $2\pi N$ radians, is the same as a 0° angle, for any integer N.

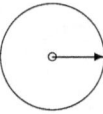

What is the size of the above angle? Here are a few of the infinite number of correct answers:

360°	0°	36,000,000°
2 $\pi\ radians$	0 $radians$	200, 000 $\pi\ radians$
-360°	720°	-36,000,000°
-2 $\pi\ radians$	4 $\pi\ radians$	-200,000 $\pi\ radians$

5.2   c, the Speed of Light

The speed of light is a physical constant, not a mathematical constant. It is not a number; it is a speed. Since it is not a number, it does not really belong in this book, but the book's author thinks it is interesting. One advantage of being the author is being able to include irrelevant sections.

The speed that light travels is represented by the symbol c. It is about 670,616,629 miles per hour.

That speed is so fast that it seems instantaneous here on earth.

Look at sunlight shining on your friend's smiling face. That light left the sun 8 minutes ago.

Look at the moon. That light left the moon 1 second ago.

In 1905 Albert Einstein proposed the famous mass-energy equation

$$E = mc^2$$

That equation says that energy (E) equals mass (m) times the speed of light squared. Because the speed of light is so large, the formula means that there is a lot of energy in a small amount of stuff (mass). That is why an atom bomb, which is fairly small, makes such a huge explosion when its matter is converted to energy.

5.3 e, Euler's Number

e is another irrational number. Its symbol is the small letter e. Its actual value is a little more than 2.71. It is the limit of this value, as n approaches infinity:

$$(1 + \frac{1}{n})^n$$

If $n = 1$, the value of the formula is 2.

If $n = 10$, the value of the formula is 2.59374246.

If $n = 100$, the value of the formula is 2.70481383.

As n gets bigger, the value of the formula gets closer to the value e, but no matter how big n gets, the value of the formula never reaches the value e.

The first part of the value of e is:

```
2.71828182845904523536028747135266249775724709369
99597496696762772407663035354759457138217852516642
74274663919320030599218174135966290435729003342952
60595630738132328627943490763233829880753195251019
```

e is also known as Euler's number. Leonhard Euler (pronounced *oiler*) was an awesome mathematician from Switzerland. He lived in the 1700's.

Ten is the base of the *common logarithm*, while e is the base of the *natural logarithm*. Intead of writing log_e, one can write ln, which means the natural logarithm. The natural logarithm and e are useful in the study of calculus. This is discussed on page 79.

The number e is remarkable in how it pops up in different branches of mathematics.

Interest Calculations Suppose you have \$1000 in your bank account. You have a very friendly bank. The bank pays you 100% interest per year. If the bank credits the interest once, then at the end of the year you have \$2000; that is called simple interest.

If the bank credits the interest every day, it is known as compounding daily interest. Every day, the bank adds interest to your bank account, giving you 1/365 of what you currently have in your bank account. With compounding daily interest, at the end of the year you would have \$2,714.57.

There is a limit to how much you would have, even if the bank compounded interest every second of every minute in every day. You would never quite have $1000 \times e$ dollars.

Probability Suppose there is a lottery in which one in every million tickets will win a pink Cadillac. If you and nine buddies each buy one million tickets, then probably you and your pals will win about ten pink Cadillacs. However, it is unlikely that each of you will win exactly one pink Cadillac; probably some will win one pink Cadillac, some will win more than one pink Cadillac, and some will win nothing at all.

Even though you bought a million tickets, there is no guarantee that you will win a pink Cadillac. Chances are better than 1 in 3 that you will win nothing. To be more precise, the probability that you will win nothing is about $1/e$, or 37%.

I wonder whether the winners will share. Who needs two pink Cadillacs?

5.4 ϕ, phi, the Golden Ratio

ϕ is a letter of the Greek alphabet, spelled *phi*, pronounced *fye*, rhymes with *by*. It represents a number that has several notable properties. It has been called the *phi-nest* number.

ϕ is the length of a *Golden Rectangle* whose width is 1. If one draws one line to divide a Golden Rectangle into a square whose sides are the width of that Golden Rectangle and a new rectangle, then the new rectangle is also a Golden Rectangle.

All Golden Rectangles have sides whose ratio is ϕ:1.

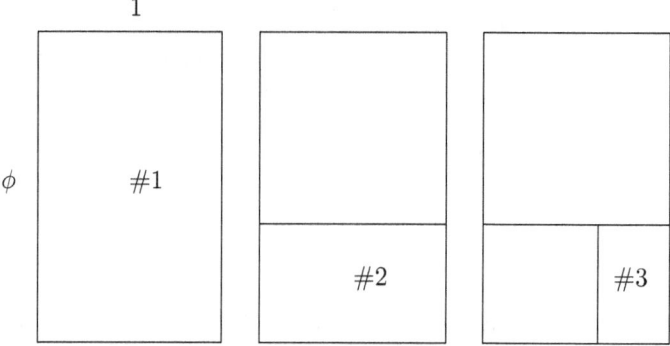

The Golden Rectangle is also considered to be aesthetically special. It's a nice shape. It's a good shape for a painting. It's a good shape for a building.

ϕ is an irrational number. Its first 199 digits are:

```
1.6180339887498948482045868343656381177203091 79805
7628621354486227052604628189024497072072041 8939113
7484754088075386891752126633862223536931793 1800607
6672635443338908659593958290563832266131992 8290267
```

The mathematical formula for ϕ can be derived from defining the width of a Golden Rectangle to be w and its length to be $w + x$. Then since Golden Rectangles have a constant ratio of their height to their width:

$$\frac{\text{height}}{\text{width}} = \frac{\phi}{1} = \frac{w+x}{w} = \frac{w}{x}$$

Since $\frac{w}{x} = \phi$ then multiplying each side of that equation by x:

$$w = x\phi$$

Substituting $x\phi$ for w above, then:

$$\phi = \frac{x\phi + x}{x\phi}$$

Factor out x:

$$\phi = \frac{\phi + 1}{\phi}$$

Multiply each side of the equation by ϕ:

$$\phi^2 = \phi + 1$$

Subtract $\phi + 1$ from both sides of the equation:

$$\phi^2 - \phi - 1 = 0$$

The *Quadratic Formula* section on page 50 can be used to solve the above equation. One solution is negative. The positive solution is:

$$\phi = \frac{1 + \sqrt{5}}{2}$$

A sweet treat for people who love the number 5:

$\phi = 5^{.5} \times .5 + .5$

5.5 i, the Imaginary Unit

i is the *imaginary unit*. Its symbol is the small letter i^1. In the complex number system, i is defined to be the square root of negative 1. Actually it is one of the two square roots of negative 1. The other is $-i$.

$$i = \sqrt{-1} \qquad i \times i = -1$$
$$-i = \sqrt{-1} \qquad -i \times -i = -1$$

There is no real number that is the square root of a negative number, because, in the real number system, a negative number times a negative number is always a positive real number. i is called imaginary because it is not part of the *real* number system. i is part of the *complex* number system. Complex numbers have a real part and an imaginary part. They have the form:

$$a + b\,i$$

where a and b are real numbers and i is the imaginary unit. The complex number system includes values that are the square roots of negative real numbers.

The mathematical terms *real* and *imaginary* can be confusing. In one sense, all numbers are imaginary: *seventeen* is the idea of a particular counting number; it is not something you can touch. In another sense, i is real: it is the idea of the value in the complex number system which is the square root of -1.

---

[1]The whiz-bang Leonhard Euler did not invent the imaginary unit, but he was the one who first used the symbol i to represent it.

Complex numbers are used in physics and electrical engineering. They tend to be good for working with waves, which have magnitude and a phase shift. Phase shift describes how far to the left or right the wave slides.

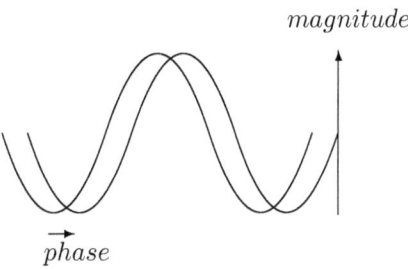

Wave and Inverted Wave

One can define a system that modifies a wave based on a complex number whose real part defines the change in magnitude and whose imaginary part defines the phase shift. In that case, i represents a phase shift, which, when multiplied twice, results in an inverted wave, because $i \times i = -1$.

Wave and Inverted Wave

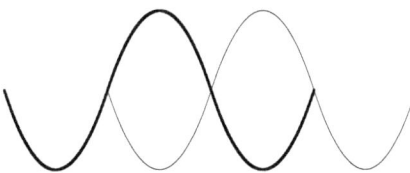

Chapter 6

Equations

An *equation* is a mathematical expression with an equal sign ($=$) in it.

6.1 Polynomial Equations

A *polynomial* is an equation constructed from variables (like x), constants (like 2), and the operations of addition, subtraction, multiplication, and exponents. This book limits itself to polynomials with only one variable, such as:

$$x^2 + 2x + 1 = 0$$

One *solves* an equation by figuring out what numbers can replace each variable, causing the equation to be true. The above equation is solved by:

$$x = -1$$

The fine number ϕ was discussed on page 44. $x = \phi$ is a solution to this polynomial equation:

$$x^2 - x - 1 = 0$$

Some polynomial equations have more than one solution:

$$x^2 + 2x - 15 = 0$$

has the solutions:

$$x = 3 \qquad\qquad x = -5$$

Quadratic Formula

A *quadratic equation* has the form

$$ax^2 + bx + c = 0, \text{ where } (a \neq 0)$$

There is a formula that allows one to solve every quadratic equation. The *quadratic formula* is:

$$x = \frac{-b \pm \sqrt{b^2 - 4ac}}{2a}$$

The symbol $\pm$ means *plus or minus*, meaning there are two solutions, one using a positive square root and the other using the negative square root.

$$x = \frac{-b + \sqrt{b^2 - 4ac}}{2a} \qquad x = \frac{-b - \sqrt{b^2 - 4ac}}{2a}$$

A lot depends on the value of $b^2 - 4ac$. If its value is:

- zero, then the two solutions are the same: $\frac{-b}{2a}$

- greater than zero, then the two solutions are real numbers

- less than zero, then the two solutions are complex numbers. This is another reason why i is such a useful number: it allows one to solve many polynomial equations.

Degrees of Polynomial Equations

Linear equations have degree 1. A linear equation has the form

$$ax + b = 0, \text{ where } (a \neq 0)$$

Quadratic equations have degree 2. The quadratic formula is used to solve quadratic equations.

Cubic equations have degree 3. A cubic equation has the form

$$ax^3 + bx^2 + cx + d = 0, \text{ where } (a \neq 0)$$

There is a formula that allows one to find all the solutions of a cubic equation.

Quartics have degree 4. There is a formula for finding all the solutions of a quartic.

Quintics have degree 5. Sadly, although quintics and above do have solutions, there is no single set of formulas that allows one to find the solutions.

6.2 Euler's Identity

$$e^{\pi i} + 1 = 0$$

The above is true. Isn't that incredible?

Why should e, Euler's number, have anything to do with π, the ratio of the circumference of a circle to its diameter? And what do either of them have to do with i, the imaginary unit?

This equation is known as Euler's identity. It is a beatiful equation. It has five of the all-time favorite numbers:

e, Euler's number, the limit of $(1 + \frac{1}{n})^n$ as n approaches infinity

π, the circumference of a circle whose diameter is 1

i, the square root of -1

1, unity. The *multiplicative identity*, because $a \times 1 = a$. One is #1. As the Greek philosopher Plato said:

>   *If one is not, then nothing is.*

0, zero. The *additive identity*, because $a + 0 = a$. A contender for the smallest number.

Euler's identity is awesome, but it is not magic; it's mathematics. The proof involves a bit of algebra, a bit of geometry, and a bit of trigonometry. Part of the proof is shown in *Euler's Formula* on page 68.

It is very satisfying to know that the above numbers are related in such an elegant way. Life is good.

Chapter 7

Various Branches of Mathematics

There are many branches of mathematics. This chapter discusses a few of them.

7.1 Topology

Topology is the area of mathematics that is concerned with topological space. Two spaces are topologically the same if one can be converted to the other without cutting or gluing. Stretching and squishing are allowed. Topologists do not care how big something is, but they do care about its edges, holes, inside, and outside.

$\bigcirc$ is topologically equivalent to $\square$ and to $\heartsuit$; each has an inside and an outside.

A hula hoop and a sewing needle are topologically equivalent; each has one hole.

A topologist is a mathematician who cannot tell the difference between a coffee cup and a donut.

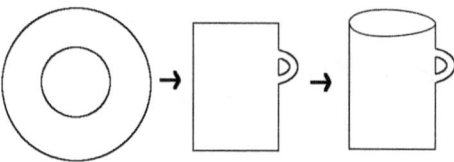

Start with a donut. Stretch it out on one side so there's a big glop of donut on the left, with the hole forming a handle on the right of the big glop. Squish the top of the big glop down into itself, forming a cup.

Möbius Strip

The Möbius strip has an interesting topology. To create a Möbius strip, take a strip of paper, give it a half-twist, and tape the ends together to form a loop with a twist in it. Draw a line along the middle of the strip. By the time you get back to where you started, both sides of the strip will have lines on them. A Möbius strip has only one side; there is no separate inside and outside.

Now cut along the line. What do you get? Two Möbius strips? No way. You don't even get two loops. Try it.

Hairy Ball Theorem

> *Hairy Ball Theorem of Topology:* You can't brush a hairy ball without creating a cowlick.

In math-speak, the hairy ball theorem states that there is no non-vanishing continuous tangent vector field on a sphere. Mathematics most certainly has its own language. I would not recommend it for social occasions.

7.2   Algebra

Algebra is the study of equations involving numbers, operations, and variables.

Question: If $x + 3 = 5$, what is the value of x?
Answer: $x = 2$.

There are many useful techniques for solving these problems. The simplest technique is to apply the same arithmetic operation to both sides of the equation. In the sample problem, one can subtract three from both sides:

$$x + 3 = 5$$
$$x + 3 - 3 = 5 - 3$$
$$x = 2$$

There are rules, such as the quadratic formula on page 50, that help one solve more complex problems, such as the equation

$$x^2 - 3x + 2 = 0$$

Algebra was easy for the Romans, because X was always 10.

Teacher: What is $7Q + 3Q$?
Student: 10 Q
Teacher: You're welcome.

7.3   Logic

Logic is the study of the principles of deduction and proof. It is
not actually a branch of mathematics, but logic and math are good
friends.

Deductive Logic

An example of deductive logic was given by Aristotle more than
2300 years ago:

1. All humans are mortal.

2. Socrates is a human.

3. Therefore, Socrates is mortal.

The above seems so obvious that even a child in kindergarden
should understand it. If it is so obvious, then what is wrong with
the below?

1. All horses have a tail.

2. All dogs have a tail.

3. Therefore, all dogs are horses.

Inductive Logic

Inductive logic makes a general conclusion based on limited obser-
vations. Use inductive logic with caution, because it can lead to
incorrect conclusions. For example, one might examine a thousand
cats and observe that each cat has a tail. Using inductive logic,
one could conclude that all cats have tails. That's not true: many
Manx cats have no tails.

Counter-Inductive Reasoning

Counter-inductive reasoning is a false form of reasoning that is based on the opposite of the principles of inductive logic. Oddly, people often use counter-inductive reasoning. For example:

> *"It has rained every day for the last ten days.*
> *It is bound to be sunny tomorrow."*

Mathematical Induction

Mathematical induction is actually a form of deductive logic, because it is a formal proof. Mathematical induction can be used to prove that a statement is true for all counting numbers 1, 2, 3, ... One only has to prove this:

1. Show it is true for the case where $n = 1$.

2. Prove that, *if* the statement is true for some value $n \geq 1$, then the statement is also true for the value $n + 1$.

Suppose you wanted to know the sum of all the integers from 1 to 1000. Mathematics has useful notation for this, using the summation symbol Σ.[1]

$$\sum_{i=1}^{3} i = 1 + 2 + 3 = 6 \qquad \sum_{i=1}^{1000} i = ?$$

We will use mathematical induction to prove, for $n \geq 1$, that

$$\sum_{i=1}^{n} i = \frac{n(n+1)}{2}$$

---

[1] Who introduced the summation symbol Σ? If your guess is the excellent Leonhard Euler, you'd be right on mark.

First, for $n = 1$, it is true that

$$\sum_{i=1}^{1} i = \frac{(1)(2)}{2} = \frac{2}{2} = 1$$

Second, if it is true that, for some value of $n \geq 1$:

$$\sum_{i=1}^{n} i = \frac{n(n+1)}{2}$$

then:

$$\sum_{i=1}^{n+1} i = (n+1) + \sum_{i=1}^{n} i$$
$$= (n+1) + \frac{n(n+1)}{2}$$
$$= \frac{2(n+1)}{2} + \frac{n(n+1)}{2}$$
$$= \frac{2(n+1) + n(n+1)}{2}$$
$$= \frac{(n+1)(2+n)}{2}$$
$$= \frac{(n+1)(n+2)}{2}$$

That is what we wanted to prove.

What is the sum of the integers from 1 to 1000? That's easy:

$$\sum_{i=1}^{1000} i = \frac{1000 \times 1001}{2} = \frac{1,001,000}{2} = 500,500$$

7.4   Geometry

Geometry is the study of shapes. One can use geometry to calculate the area of a rectangle (*width* × *height*) or a circle ($\pi\, r^2$, where r is half the diameter).

Solid geometry is used to calculate the volume of a solid. This is useful if Farmer Jo wants to know how much sunflower seed can be put into a storage bin.

In *Euclidean geometry*, two lines that are parallel will never intersect.

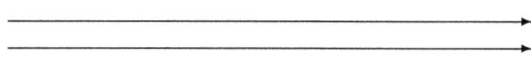

Non-Euclidean geometry can be interesting: parallel lines might intersect. Think about the geometry of the surface of a sphere such as Earth. Latitude lines, which are lines parallel to the equator, do not touch; however, longitude lines, which go from pole to pole, are parallel at the equator, but meet at the poles.

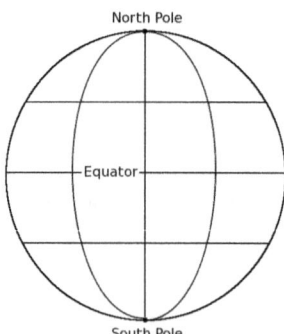

Cartesian Coordinates

Many branches of mathematics, including geometry, use cartesian coordinates, with a horizontal x-axis and a vertical y-axis. A point is defined as a pair (x, y).

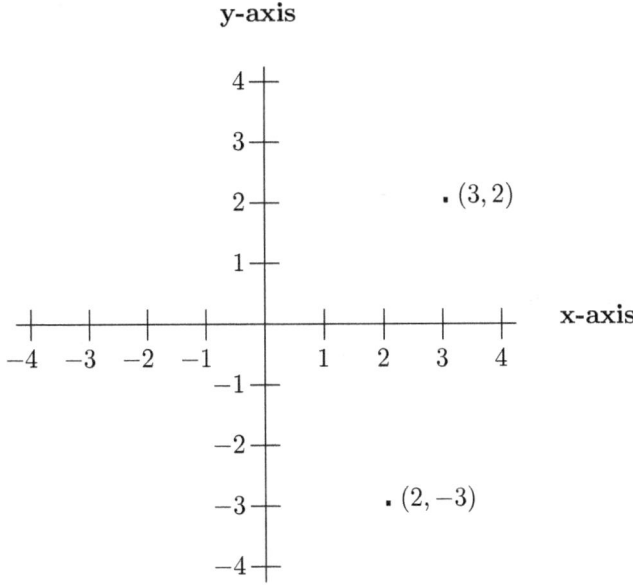

The *complex plane*, also known as the *imaginary plane*, uses cartesian coordinates to represent complex numbers. In that case, the y-axis is called *the imaginary axis*, and the point (2, -3) represents the complex number $2 - 3i$.

Pythagorean Theorem

The section *Square Root of Two* on page 30 noted that $\sqrt{2}$ is the length of the diagonal line in a square whose area is 1. That fact is supported by the Pythagorean theorem. Pythagoras was a Greek mathematician who lived around 500 BC.

A *right triangle* is a triangle having a right angle. A *right angle* is a 90° angle.

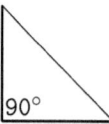

Sometimes a small box is drawn in the 90° corner of a right triangle.

The Pythagorean Theorem asserts: *for any right triangle, the square of the longest side is equal to the sum of the squares of the other two sides.*

$$c^2 = a^2 + b^2$$

We had already learned that $c = \sqrt{2}$ in the case where $a = b = 1$. The Pythagorean theorem holds true for all right triangles.

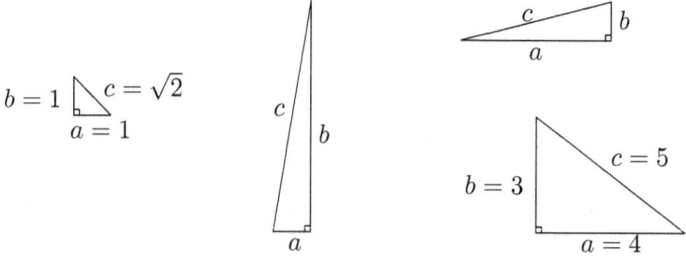

There are many proofs of the Pythagorean Theorem. Regrettably, none are in this book.

7.5 Trigonometry

Trigonometry is a branch of mathematics that deals with right triangles.

Basic Trigonometric Functions

Trigonometry defines three basic functions, *sine* (sin), *cosine* (cos), and *tangent* (tan). Each of these functions is concerned with the ratio of the lengths of two sides of a right triangle. Each function shows how the ratio changes as the other angles of the right triangle change.

The character θ is called *theta*; it is a letter of the Greek alphabet. By convention, it is often used as a variable representing an angle.

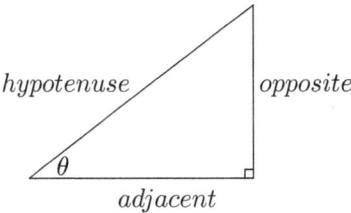

The side of the triangle opposite the right angle is called the **hypotenuse**.

The side of the triangle opposite θ is called the **opposite** side.

The other side of the triangle, between θ and the right angle, is called the **adjacent** side.

Each of the three basic trigonometric functions defines a ratio of the lengths of two of the sides of a right triangle.

function	ratio	mnemonic (memory aid)
$\sin(\theta)$	**O**pposite / **H**ypotenuse	***S**addle **O**ur **H**orse.*
$\cos(\theta)$	**A**djacent / **H**ypotenuse	***C**anter **A**way **H**appily*
$\tan(\theta)$	**O**pposite / **A**djacent	***T**o **O**ld **A**untie.*

If the length of the hypotenuse is 1, then the other two sides have lengths of $sin(\theta)$ and $cos(\theta)$.

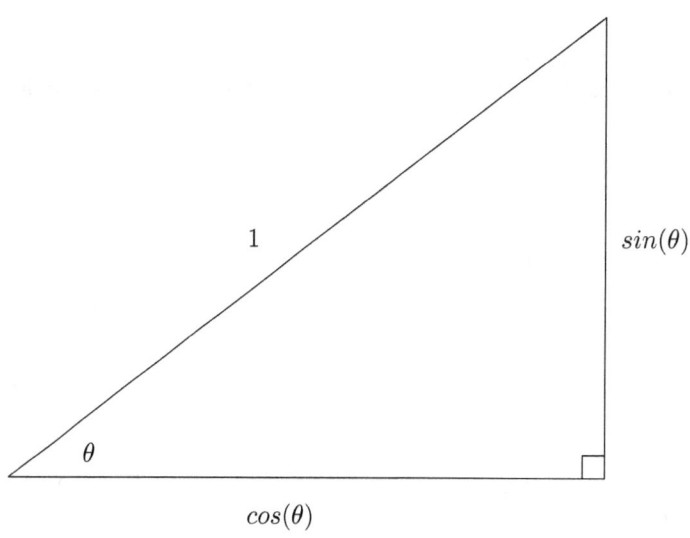

One can view the right triangles as being placed inside a unit circle, which is a circle whose radius is 1. The triangles have a hypotenuse whose length is 1.

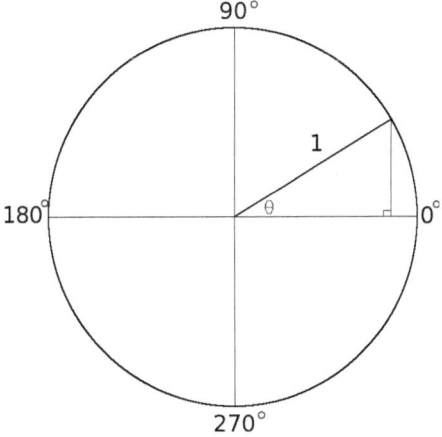

Right Triangle, $\theta = 31°$

There is no triangle when the angle is straight up or straight across. The horizontal lines at $0°$ and $180°$ have *opposite* $= 0$, while the vertical lines at $90°$ and $270°$ have *adjacent* $= 0$.

Fact: As long as trigonometry is taught in school, there will be prayer in school.

As θ gets smaller and approaches zero, the triangle gets very short and flat: the opposite side gets very small and the adjacent side gets almost as long as the hypotenuse. So, as θ approaches zero, the value of the *sin* function approaches zero, while the value of the *cos* function approaches one.

On the other hand, as θ gets close to 90°, 0.5π radians, the triangle gets very narrow and tall: the adjacent side gets very small and the oposite side gets almost as long as the hypotenuse. So, as θ approaches 90°, the value of the *sin* function approaches one, while the value of the *cos* function approaches zero.

$\theta = 7°$
$\theta = .12\,\text{radians}$ $\theta = 81°$
                                                            $\theta = .45\pi\,\text{radians}$

The graphs of the sine and cosine functions are alike, except that they are shifted 90°, $\pi/2$ radians, along the x-axis. These functions usually expect their argument to be in radians, not degrees. Their curve repeats every 2π radians, because adding or subtracting 2π radians does not change an angle; a 0 radian angle is the same as a 2π, 4π, 6π, ... radian angle.

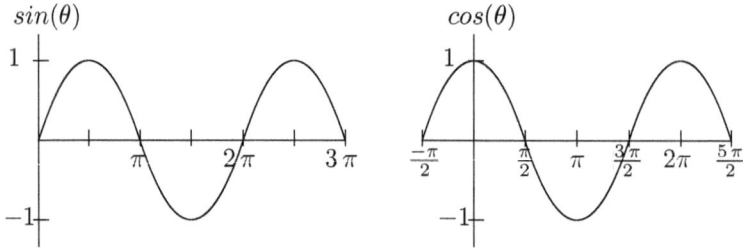

Complex Numbers in Trigonometric Form

Every complex number $a + bi$ has a corresponding trigonometric form:
$$h\left(\cos(\theta) + i\,\sin(\theta)\right)$$
where h and θ are defined as follows:

1. Map the complex number $a + bi$ to a point (a, b) on the complex plane.

2. Create a right triangle whose hypotenuse extends from (0,0) to (a, b), with the vertical side going from (a,b) to the real axis.

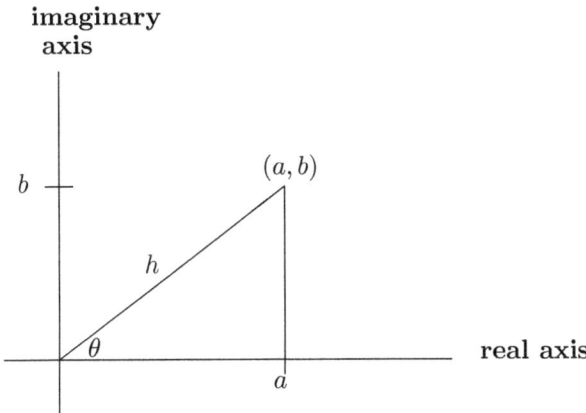

3. θ is the angle at the point (0,0).

4. h is the length of the hypotenuse. Using the Pythagorean theorem:
$$h = \sqrt{a^2 + b^2}$$

Euler's Formula

$$e^{xi} = \cos(x) + i\,\sin(x)$$

Euler's formula is an equation for raising the number e to an imaginary power. In the formula, x is an angle measured in radians.

This book does not provide a complete proof of Euler's formula. One proof involves any complex number z on the circumference of the unit circle in the complex plane. Map z to its trigonometric form:

$$z = \cos(\theta) + i\,\sin(\theta)$$

Differential calculus mumbo jumbo can then be used to show that $\log_e(z) = \ln(z) = \theta\,i$. That means that $z = e^{\theta i}$, so:

$$e^{\theta i} = \cos(\theta) + i\,\sin(\theta)$$

If we presume that Euler's formula is true, we can use it to show that Euler's identity (page 52) is true, in addition to being beautiful.

Note that:

$$\sin(\pi) = 0 \qquad \cos(\pi) = -1$$

Using Euler's formula:

$$e^{\pi i} = \cos(\pi) + i\,\sin(\pi)$$
$$e^{\pi i} = -1 + (i \times 0)$$
$$e^{\pi i} = -1$$

Adding 1 to each side of the equation results in Euler's identity:

$$e^{\pi i} + 1 = 0$$

Question: How many mathematicians does it take to change a light bulb?
Answer: $-e^{\pi i}$. In other words, 1.

7.6 Probability Theory

Probability theory is the branch of mathematics that analyzes random events. It is used in statistics, when analyzing large amounts of data. It is also used by gamblers.

Probability values range between zero and one. A probability of zero means the event will never happen, while a probability of one or 100% means the event is guaranteed to occur. An important principle is that if the chance for one event occurring is A and the chance for a separate event occurring is B, then the probability that both events will occur is $A \times B$, not $A + B$.

Rolling Dice

One can calculate the probability of particular results when rolling a pair of dice.

doubles $\frac{1}{6}$, the odds that the second die will match the first.

seven $\frac{1}{6}$, the odds that the second die will show a value that, when added to the value of the first die, equals seven.

eleven $\frac{1}{3} \times \frac{1}{6} = \frac{1}{18}$, the odds that the first die will be 5 or 6, times the odds that the second die will have a value that, when added to the value of the first die, equals eleven.

snake-eyes $\frac{1}{6} \times \frac{1}{6} = \frac{1}{36}$, the odds that both dice will display 1.

Make some money: Bet your friends that they won't roll snake-eyes. If they roll snake-eyes, you will give them a quarter. If they don't, they must pay you a penny. On average, for every 36 rolls of the dice, you will pay out 25 cents and receive 35 cents.

But don't quit school yet. There are better ways to make money.

Birthday Coincidence

The *birthday coincidence* states that, for a random group of twenty-three people, chances are about 50-50 that two people in the group have the same birthday.

Explanation

To simply the explanation, presume there are always 365 days in a year.

The probability that two random people have different birthdays is the probability that the second person does not have the same birthday as the first person:

$$\frac{364}{365} \approx 0.9972$$

The probability that three random people have different birthdays is the probability above, times the probability that the third person does not have the same birthday as either of the first two people. This is:

$$\frac{364}{365} \times \frac{363}{365} \approx 0.9918$$

The probability that twenty-three random people have different birthdays is the probability that twenty-two random people have different birthdays, times the probability that the twenty-third person does not have the same birthday as any of the first twenty-two people. This is:

$$\frac{364}{365} \times \frac{363}{365} \times \ldots \times \frac{344}{365} \times \frac{343}{365} \approx 0.49$$

So, chances are a little better than fifty percent that two people in the group of 23 people have the same birthday.

Three mathematicians who specialize in probability theory go hunting. When they see a rabbit, the first one shoots, but misses on the left. The second one shoots and misses on the right. The third one shouts: "We hit it!"

A statistician has his head in the oven while his feet are frozen inside a huge chunk of ice. He says, "On average, I feel just fine."

7.7 Calculus

Calculus is the study of functions and limits. Calculus has two main branches: integral calculus and differential calculus.

Integral Calculus

Integral calculus is the study of integration. *Integration* involves determining the *integral*, which can be envisioned as the area under the graph of a curve.

Since ancient times, mathematicians and scientists have wanted to be able to calculate the area under a curve. That could be used, for example, to calculate how much beer a barrel would hold.

Suppose the curve was defined by the equation $y = x^2$. This is known as *the function of x squared*, and can be written $f(x) = x^2$. Suppose the area to be calculated was bounded by $x = 1$ and $x = 3$.

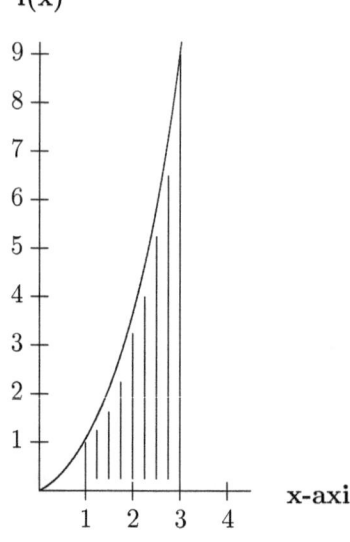

One method of approximating the area is to divide it into rectangles whose bottom is the x-axis and whose top intersects the graph of the function. The area of a rectangle is easy to calculate: width times height. The sum of the areas of the rectangles is a good approximation of the area under the curve. As the width of the rectangles gets smaller, there are more rectangles and the approximation becomes more accurate. The width of the rectangle is *the difference in x*, also known as d*x*.

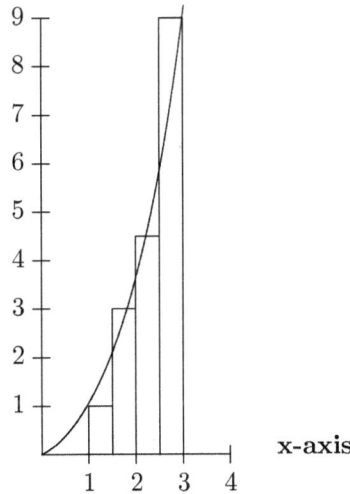

The sum of the areas of the rectangles above is:

$$.5 + 1.5 + 2.25 + 4.5 = 8.75 = 8\frac{3}{4}$$

The exact value of the integral will be calculated later, on page 80 in the section on the *Fundamental Theorem of Calculus*.

Integral Notation

The notation for integral calculus is strange-looking. Using the notation of integral calculus, we have estimated the value of

$$\int_1^3 x^2 \, \mathrm{d}x$$

to be approximately $8\frac{3}{4}$. The above is pronounced *the integral from 1 to 3 of the function x^2 with respect to x.*

More general notation looks like this:

$$\int_a^b f(x) \, \mathrm{d}x$$

and is pronounced *the integral from* a *to* b *of a function of x with respect to x.*

The notation $\int_a^b \ldots \mathrm{d}x$ represents the limit of the sums of the areas of the rectangles between the vertical lines at $x = a$ and $x = b$, as the width of the rectangles gets smaller and more rectangles are formed.

The notation $f(x) \, \mathrm{d}x$ can be thought of as the area of one rectangle: its height, $f(x)$, times its width, $\mathrm{d}x$. By the way, the notation $f(x)$ to represent *a function of x* was invented by ... you guessed it: Leonhard Euler.

The integration symbol $\int$ evolved from the letter S and represents the *sum* of the areas of the rectangles.

Differential Calculus

Differential calculus is the study of the slope of functions. On the graph of the function at a particular point, the *derivative* can be thought of as the ratio of the change in the y value corresponding to a small change in the x value:

$$\frac{dy}{dx}$$

Differentiation means calculating a derivative. For linear functions, whose graph is a straight line, the derivative is the same at any point. A function whose graph is a horizontal line, such as $f(x) = 2$, has a derivative of 0. The function $f(x) = x$ has a derivative of 1.

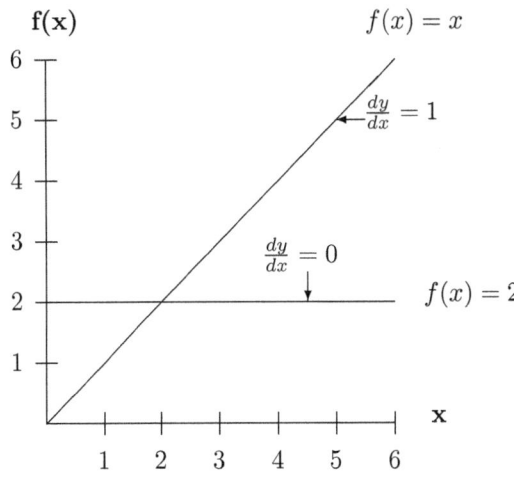

The derivatives of non-linear functions are not constants, but there are rules that help one to compute these derivatives. One helpful rule is:

$$\frac{d}{dx}x^n = nx^{n-1}$$

Consider the non-linear function $f(x) = x^2$. The above rule states its derivative is $2x$.

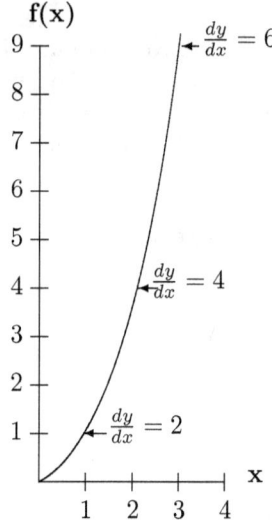

A related concept is the *antiderivative*. An *antiderivative* of a function f is a function F whose derivative is equal to f. For example, the function $f(x) = 2x$ has an antiderivative function $F(x) = x^2$.

Distance, Speed, and Acceleration

Imagine a spaceship. The ship is currently parked at the Earth Space Station. You are the officer in charge of acceleration. You have three meters.

| | 0 | miles |

The *odometer* measures distance traveled in miles.

| | 0 | MPS, miles per second |

The *speedometer* measures current speed in MPS, miles per second. Speed is the derivative of distance traveled with respect to time.

| | 0 | MPSS, miles per second per second |

The *accelerometer* measures current acceleration in MPSS, the change in MPS every second. Acceleration is the derivative of speed with respect to time. If the acceleration is zero, the speed is not changing. If the acceleration is greater than zero, you are speeding up. If the acceleration is less than zero, you are slowing down.

The captain has decided to have lunch on the moon, about 230,000 miles from Earth. It is time to take off. You set the acceleration rate to 2 MPSS; this means the speed will increase by 2 MPS every second.

time	odometer	speedometer	accelerometer
t	t^2	$2t$	
1 sec.	1 mile	2 MPS	2 MPSS
5 sec.	25 miles	10 MPS	2 MPSS
60 sec.	3,600 miles	120 MPS	0 MPSS

One minute after takeoff, you have gone 3,600 miles and your speed is 120 miles per second (432,000 MPH). That is your target speed for this short jaunt, so you reset the acceleration rate to 0 MPSS.

After about half an hour, you are 3,600 miles from the moon. It is time to slow down, so you set the acceleration rate to -2 MPSS. One minute after that, the ship stops on the moon.

Piece of cake.

e, Euler's Number (again)

Derivative of e^x

Perhaps the most wonderful property of the number e is that the derivative of e^x is itself, e^x. This can be written:

$$\frac{d}{dx}e^x = e^x$$

No other non-zero number has the above property. The derivative of n^x is less than n^x when n is between 0 and e, and the derivative is greater than n^x when n is greater than e.

This makes e a very important number, since it is eee-z to calculate the derivative and the antiderivative of e^x — the values are e^x.

Derivative of $ln(x)$

The function that is the natural logarithm of x, $log_e(x)$, also known as $ln(x)$, has a derivative of $1/x$:

$$\frac{d}{dx}(ln(x)) = \frac{1}{x}$$

The derivatives of logarithms to bases that are not e are more difficult to calculate. This makes natural logarithms extremely useful, since it is eee-z to calculate their derivatives. In fact, it is this simplicity that is the reason that a logarithm to the base e is called the *natural* logarithm.

Fundamental Theorem of Calculus

Until the fundamental theorem of calculus appeared, integral calculus and differential calculus were viewed as two separate areas of mathematics. The *fundamental theorem of calculus* states that integration is the inverse of differentiation. This was a major mathematical breakthrough, since it allows one to calculate integrals by using derivatives. There are many useful rules for calculating derivatives, so that makes it easier to calculate integrals.

The fundamental theorem is:

$$\int_a^b f(x)\,dx = F(b) - F(a)$$

where F is an antiderivative function of f. Remember the example for integral calculus on page 72? It was:

$$\int_1^3 x^2\,dx$$

An antiderivative of the function $f(x) = x^2$ is the function

$$F(x) = \frac{x^3}{3}$$

Using the fundamental theorem of calculus,

$$\int_1^3 x^2\,dx = F(3) - F(1) = \frac{27}{3} - \frac{1}{3} = \frac{26}{3} = 8\frac{2}{3}$$

We were not very far off when we used the method of approximating the integral by adding the areas of rectangles, which came out to $8\frac{3}{4}$. Still, it is better to be exact, particularly if one is doing the calculations for a mission to Mars.

7.8   Transfinite Arithmetic

One hundred fifty years ago, people presumed that infinity was absolute. It was not mathematically interesting. Then Georg Cantor arrived on the scene. He proved that there are an infinity of different infinities, each one larger than the one before. The study of these numbers is called *transfinite arithmetic*.

Cantor described the idea of *cardinal numbers*, which measure the size of a set. Two sets are the same size if one can define a one-to-one map from each set to the other. For example, if there is a pile of apples here and a pile of teachers there, and if there is an apple for each teacher and a teacher for each apple, then the two piles have the same cardinality, or size. However, if there is an apple for each teacher, but not a teacher for each apple, then there are more apples than teachers.

Infinite numbers follow different rules than finite numbers. Would you believe there are as many counting numbers as there are even counting numbers? That is true, because for each counting number N, one can define a corresponding even counting number (N + N), and also for each even counting number M, one can define a unique counting number (M / 2). The set of counting numbers and the set of even counting numbers have the same cardinality.

numbers	even numbers
1	2
2	4
3	6
. . .	. . .

ℵ is the first letter of the Hebrew alphabet. $\aleph_0$ is called aleph-zero or aleph-null. It is the smallest infinite cardinal number. It is the countable infinity. It is the size of the set of counting numbers. It is the size of the set of even counting numbers. It is also the same as the number of integers and the number of rational numbers. In fact,

$$\aleph_0 + \aleph_0 = \aleph_0$$

$$\aleph_0 \times \aleph_0 = \aleph_0$$

However, $\aleph_0$ is *not* the same as the number of real numbers.

Diagonal Argument

There is another infinite number that is larger than $\aleph_0$. It is the number of real numbers. In fact, there are more real numbers between zero and one than there are counting numbers.

Cantor proved this in a rather elegant way. It is called his *diagonal argument*. He said we should try to imagine an infinite ordered list of *all* the real numbers between zero and one. If we could make such a list, then the number of real numbers between zero and one would be $\aleph_0$, because the list would be a 1:1 map between the counting numbers and the real numbers between zero and one.

Cantor's proof shows that no ordered list of real numbers between zero and one can contain *all* the real numbers between zero and one. Consider Table 7.1.

Table 7.1: Infinite List of Numbers Between 0 and 1

	1st digit	2nd digit	3rd digit	4th digit	
#1	**.2**	3	1	2	121212...
#2	.4	**4**	4	4	4444...
#3	.0	0	**0**	9	000...
...	...	...	...	...	...

The number that is missing from Table 7.1 has these properties:

- The 1st digit after the decimal point is not the same as the 1st digit after the decimal point for #1 in Table 7.1. This means the 1st digit of the missing number is not 2.

- The 2nd digit after the decimal point is not the same as the 2nd digit after the decimal point for #2 in Table 7.1. This means the 2nd digit of the missing number is not 4.

- The 3rd digit after the decimal point is not the same as the 3rd digit after the decimal point for #3 in Table 7.1. This means the 3rd digit of the missing number is not 0.

- The nth digit after the decimal point is not the same as the nth digit after the decimal point for #n in Table 7.1.

If someone claims the *Missing Number* is in fact on the list, ask which # position it has. If they say it is at position #1000, you can explain it is not there, because the 1000th digit after the decimal point in the *Missing Number* is not the same as the 1000th digit after the decimal point for #1000.

Therefore, any ordered list of real numbers between zero and one does not contain *all* the real numbers between zero and one, so there are more real numbers between zero and one than there are counting numbers. Q.E.D.[2]

---

[2]This is a Latin abbreviation for *quod erat demonstrandum*, which means *that which was to be demonstrated*; it means the proof has been completed.

Transfinite Numbers

Cantor also showed that there are an infinite number of infinite numbers. He called these $\aleph_0$, $\aleph_1$, $\aleph_2$, ... Each is larger than the one before. The set of *cardinal numbers* consists of these transfinite numbers, plus the counting numbers. Cardinal numbers can be used to measure the size of a set, whether the set is finite, like the number of pennies in my pocket, or infinite, like the number of counting numbers.

Infinity is not just a black hole of forever; it is mathematically interesting. Q.E.D.

7.9   Chaos Theory

Chaos theory is a great term. The term sounds like an oxymoron (a contradiction). How can one formally study wild craziness? How can one predict unpredictable behavior?

The idea is that there is wild crazy behavior, like weather, which seems random but is not. It seems random because the causes are so complex that we are unable to predict the behavior. The chaos that is of interest is also called a *dynamic system*. It has these characteristics:

- sensitive to initial conditions

- small causes can have big effects

Butterfly Effect

Chaos theory has come up with some marvelous descriptions. The *butterfly effect* is a term that arose from the title of a presentation in 1972: *Predictability: Does the Flap of a Butterfly's Wings in Brazil set off a Tornado in Texas?* The butterfly effect is what makes it impossible to accurately predict next week's weather.

Rogue Waves

Rogue waves are freak waves that unexpectedly appear in the ocean. A rogue wave can sink a huge ship on a calm day. Chaos theory may help explain how these waves occur. The ocean may be as chaotic as the atmosphere, and if a butterfly can cause a distant tornado, then perhaps a diving dolphin can cause a distant rogue wave.

Fractals

Fractals are related to chaos theory. A *fractal* is an image that is not a simple geometric shape and which can be split into smaller images that resemble the whole image. Fractals are found in nature: snowflakes, ferns, seashells. Other fractals have mathematical definitions. If you zoom in on a mathematical fractal, a section has as much detail as the whole fractal.

Fractal Image of a Mandelbrot Set

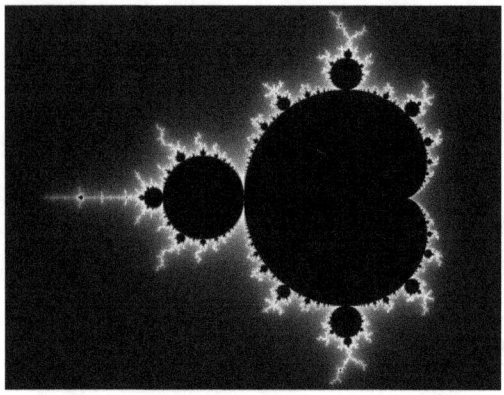

The equation that is used to define a Mandelbrot set is

$$z_{n+1} = z_n{}^2 + c$$

That equation defines a series of numbers, z_0, z_1, ... Start with $z_0 = 0$. A Mandelbrot set is the set of complex numbers c for which the series is bounded; i.e., the series does not approach infinity.

For example, when $c = 2$ the series is:

$$0$$
$$2$$
$$6$$
$$38$$
$$1,446$$
$$\ldots$$

Since those values keep increasing, 2 is not in the Mandelbrot set.

When c is the imaginary unit i, the series is:

$$0$$
$$i$$
$$-1 + i$$
$$-i$$
$$-1 + i$$
$$-i$$
$$-1 + i$$
$$-i$$
$$-1 + i$$
$$-i$$
$$-1 + i$$
$$-i$$
$$\ldots$$

Since the sequence values keeps repeating, i is in the Mandelbrot set. When the Mandelbrot set is graphed on the complex plane, its edges form a fractal.

The term *fractal* was coined by Benoît Mandelbrot. In 1982 he published a book entitled *The Fractal Geometry of Nature*. The book has fabulous illustrations of fractals.

Mandelbrot wrote in the introduction to the book:

> "Clouds are not spheres,
> mountains are not cones,
> coastlines are not circles,
> and bark is not smooth,
> nor does lightning travel in a straight line."

Mathematics is not as simple as arithmetic, and it is far more interesting than arithmetic. We started with counting numbers. Fractal geometry is a good place to end.

Index